THIS BLOG PLANNER BELONGS TO

_____

_____

_____

CONTACT DETAILS

_____

_____

_____

_____

Dedication

This Blog Planner Book is dedicated to all the bloggers out there who want to map out their blog post ideas and document their findings in the process.

You are my inspiration for producing books and I'm honored to be a part of keeping all of your blog notes and records organized.

This journal notebook will help you record the details about your planning your blog post.

Thoughtfully put together with these sections to record: Mind mapping, Subject, Ideas, Rough Draft, Pictures & Graphics, and Notes.

How to Use this Book

The purpose of this book is to keep all of your Blog Planning notes all in one place. It will help keep you organized.

This Blog Planner Book will allow you to accurately document every detail about your blog post planning.

Here are examples of the prompts for you to fill in and write about your experience in this book:

1. Mind Mapping - Mind mapping circles for brainstorming.

2. Subject - Write the subject of your blog post.

3. Ideas - Record your ideas for your blog post.

4. Rough Draft - Write your rough draft copy.

5. Pictures & Graphics - Log what pictures and graphics you want to use.

6. Notes - Log any additional important information you want.

MY BLOG IDEAS
MIND MAP

MY BLOG IDEAS

SUBJECT

IDEAS

PICTURES/GRAPHICS

ROUGH DRAFT

NOTES

OTHER

MY BLOG IDEAS
MIND MAP

MY BLOG IDEAS

SUBJECT

IDEAS

PICTURES/GRAPHICS

ROUGH DRAFT

NOTES

OTHER

MY BLOG IDEAS
MIND MAP

MY BLOG IDEAS

SUBJECT

_____
_____
_____
_____
_____

IDEAS

_____
_____
_____
_____
_____

PICTURES/GRAPHICS

ROUGH DRAFT

NOTES

_____
_____
_____
_____

OTHER

_____
_____
_____
_____

MY BLOG IDEAS
MIND MAP

MY BLOG IDEAS

SUBJECT

IDEAS

PICTURES/GRAPHICS

ROUGH DRAFT

NOTES

OTHER

MY BLOG IDEAS
MIND MAP

MY BLOG IDEAS

SUBJECT

IDEAS

PICTURES/GRAPHICS

ROUGH DRAFT

NOTES

OTHER

MY BLOG IDEAS
MIND MAP

MY BLOG IDEAS

SUBJECT

IDEAS

PICTURES/GRAPHICS

ROUGH DRAFT

NOTES

OTHER

MY BLOG IDEAS
MIND MAP

MY BLOG IDEAS

SUBJECT

IDEAS

PICTURES/GRAPHICS

ROUGH DRAFT

NOTES

OTHER

MY BLOG IDEAS
MIND MAP

MY BLOG IDEAS

SUBJECT

_____
_____
_____
_____
_____
_____

IDEAS

_____
_____
_____
_____
_____
_____

PICTURES/GRAPHICS

ROUGH DRAFT

NOTES

_____
_____
_____
_____

OTHER

_____
_____
_____
_____

MY BLOG IDEAS
MIND MAP

MY BLOG IDEAS

SUBJECT

IDEAS

PICTURES/GRAPHICS

ROUGH DRAFT

NOTES

OTHER

MY BLOG IDEAS
MIND MAP

MY BLOG IDEAS

SUBJECT

_____
_____
_____
_____
_____
_____

IDEAS

_____
_____
_____
_____
_____
_____

PICTURES/GRAPHICS

ROUGH DRAFT

NOTES

_____
_____
_____
_____
_____

OTHER

_____
_____
_____
_____
_____

MY BLOG IDEAS
MIND MAP

MY BLOG IDEAS

SUBJECT

_____
_____
_____
_____
_____

IDEAS

_____
_____
_____
_____
_____

PICTURES/GRAPHICS

ROUGH DRAFT

NOTES

_____
_____
_____
_____

OTHER

_____
_____
_____
_____

MY BLOG IDEAS
MIND MAP

MY BLOG IDEAS

SUBJECT

IDEAS

PICTURES/GRAPHICS

ROUGH DRAFT

NOTES

OTHER

MY BLOG IDEAS
MIND MAP

MY BLOG IDEAS

SUBJECT

IDEAS

PICTURES/GRAPHICS

ROUGH DRAFT

NOTES

OTHER

MY BLOG IDEAS
MIND MAP

MY BLOG IDEAS

SUBJECT

_____
_____
_____
_____
_____

IDEAS

_____
_____
_____
_____
_____

PICTURES/GRAPHICS

ROUGH DRAFT

NOTES

_____
_____
_____
_____

OTHER

_____
_____
_____
_____

MY BLOG IDEAS
MIND MAP

MY BLOG IDEAS

SUBJECT

IDEAS

PICTURES/GRAPHICS

ROUGH DRAFT

NOTES

OTHER

MY BLOG IDEAS
MIND MAP

MY BLOG IDEAS

SUBJECT

_____
_____
_____
_____
_____
_____

IDEAS

_____
_____
_____
_____
_____
_____

PICTURES/GRAPHICS

ROUGH DRAFT

NOTES

_____
_____
_____
_____
_____

OTHER

_____
_____
_____
_____
_____

MY BLOG IDEAS
MIND MAP

MY BLOG IDEAS

SUBJECT

_____
_____
_____
_____
_____
_____

IDEAS

_____
_____
_____
_____
_____
_____

PICTURES/GRAPHICS

ROUGH DRAFT

NOTES

_____
_____
_____
_____
_____

OTHER

_____
_____
_____
_____
_____

MY BLOG IDEAS
MIND MAP

MY BLOG IDEAS

SUBJECT

IDEAS

PICTURES/GRAPHICS

ROUGH DRAFT

NOTES

OTHER

MY BLOG IDEAS
MIND MAP

MY BLOG IDEAS

SUBJECT

IDEAS

PICTURES/GRAPHICS

ROUGH DRAFT

NOTES

OTHER

MY BLOG IDEAS
MIND MAP

MY BLOG IDEAS

SUBJECT

IDEAS

PICTURES/GRAPHICS

ROUGH DRAFT

NOTES

OTHER

MY BLOG IDEAS
MIND MAP

MY BLOG IDEAS

SUBJECT

_____
_____
_____
_____
_____

IDEAS

_____
_____
_____
_____
_____

PICTURES/GRAPHICS

ROUGH DRAFT

NOTES

_____
_____
_____
_____

OTHER

_____
_____
_____
_____

MY BLOG IDEAS
MIND MAP

MY BLOG IDEAS

SUBJECT

IDEAS

PICTURES/GRAPHICS

ROUGH DRAFT

NOTES

OTHER

MY BLOG IDEAS
MIND MAP

MY BLOG IDEAS

SUBJECT

_____
_____
_____
_____
_____
_____

IDEAS

_____
_____
_____
_____
_____
_____

PICTURES/GRAPHICS

ROUGH DRAFT

NOTES

_____
_____
_____
_____

OTHER

_____
_____
_____
_____

MY BLOG IDEAS
MIND MAP

MY BLOG IDEAS

SUBJECT

IDEAS

PICTURES/GRAPHICS

ROUGH DRAFT

NOTES

OTHER

MY BLOG IDEAS
MIND MAP

MY BLOG IDEAS

SUBJECT

IDEAS

PICTURES/GRAPHICS

ROUGH DRAFT

NOTES

OTHER

MY BLOG IDEAS
MIND MAP

MY BLOG IDEAS

SUBJECT

IDEAS

PICTURES/GRAPHICS

ROUGH DRAFT

NOTES

OTHER

MY BLOG IDEAS
MIND MAP

MY BLOG IDEAS

SUBJECT

_____
_____
_____
_____
_____
_____

IDEAS

_____
_____
_____
_____
_____
_____

PICTURES/GRAPHICS

ROUGH DRAFT

NOTES

_____
_____
_____
_____

OTHER

_____
_____
_____
_____

MY BLOG IDEAS
MIND MAP

MY BLOG IDEAS

SUBJECT

IDEAS

PICTURES/GRAPHICS

ROUGH DRAFT

NOTES

OTHER

MY BLOG IDEAS
MIND MAP

MY BLOG IDEAS

SUBJECT

IDEAS

PICTURES/GRAPHICS

ROUGH DRAFT

NOTES

OTHER

MY BLOG IDEAS
MIND MAP

MY BLOG IDEAS

SUBJECT

IDEAS

PICTURES/GRAPHICS

ROUGH DRAFT

NOTES

OTHER

MY BLOG IDEAS
MIND MAP

MY BLOG IDEAS

SUBJECT

IDEAS

PICTURES/GRAPHICS

ROUGH DRAFT

NOTES

OTHER

MY BLOG IDEAS
MIND MAP

MY BLOG IDEAS

SUBJECT

IDEAS

PICTURES/GRAPHICS

ROUGH DRAFT

NOTES

OTHER

MY BLOG IDEAS
MIND MAP

MY BLOG IDEAS

SUBJECT

IDEAS

PICTURES/GRAPHICS

ROUGH DRAFT

NOTES

OTHER

MY BLOG IDEAS
MIND MAP

MY BLOG IDEAS

SUBJECT

IDEAS

PICTURES/GRAPHICS

ROUGH DRAFT

NOTES

OTHER

MY BLOG IDEAS
MIND MAP

MY BLOG IDEAS

SUBJECT

IDEAS

PICTURES/GRAPHICS

ROUGH DRAFT

NOTES

OTHER

MY BLOG IDEAS
MIND MAP

MY BLOG IDEAS

SUBJECT

IDEAS

PICTURES/GRAPHICS

ROUGH DRAFT

NOTES

OTHER

MY BLOG IDEAS
MIND MAP

MY BLOG IDEAS

SUBJECT

IDEAS

PICTURES/GRAPHICS

ROUGH DRAFT

NOTES

OTHER

MY BLOG IDEAS
MIND MAP

MY BLOG IDEAS

SUBJECT

IDEAS

PICTURES/GRAPHICS

ROUGH DRAFT

NOTES

OTHER

MY BLOG IDEAS
MIND MAP

MY BLOG IDEAS

SUBJECT

IDEAS

PICTURES/GRAPHICS

ROUGH DRAFT

NOTES

OTHER

MY BLOG IDEAS
MIND MAP

MY BLOG IDEAS

SUBJECT

IDEAS

PICTURES/GRAPHICS

ROUGH DRAFT

NOTES

OTHER

MY BLOG IDEAS
MIND MAP

MY BLOG IDEAS

SUBJECT

IDEAS

PICTURES/GRAPHICS

ROUGH DRAFT

NOTES

OTHER

MY BLOG IDEAS
MIND MAP

MY BLOG IDEAS

SUBJECT

_____
_____
_____
_____
_____
_____

IDEAS

_____
_____
_____
_____
_____
_____

PICTURES/GRAPHICS

ROUGH DRAFT

NOTES

_____
_____
_____
_____
_____

OTHER

_____
_____
_____
_____
_____

MY BLOG IDEAS
MIND MAP

MY BLOG IDEAS

SUBJECT

_____
_____
_____
_____
_____
_____

IDEAS

_____
_____
_____
_____
_____
_____

PICTURES/GRAPHICS

ROUGH DRAFT

NOTES

_____
_____
_____
_____

OTHER

_____
_____
_____
_____

MY BLOG IDEAS
MIND MAP

MY BLOG IDEAS

SUBJECT

IDEAS

PICTURES/GRAPHICS

ROUGH DRAFT

NOTES

OTHER

MY BLOG IDEAS
MIND MAP

MY BLOG IDEAS

SUBJECT

IDEAS

PICTURES/GRAPHICS

ROUGH DRAFT

NOTES

OTHER

MY BLOG IDEAS
MIND MAP

MY BLOG IDEAS

SUBJECT

IDEAS

PICTURES/GRAPHICS

ROUGH DRAFT

NOTES

OTHER

MY BLOG IDEAS
MIND MAP

MY BLOG IDEAS

SUBJECT

IDEAS

PICTURES/GRAPHICS

ROUGH DRAFT

NOTES

OTHER

MY BLOG IDEAS
MIND MAP

MY BLOG IDEAS

SUBJECT

IDEAS

PICTURES/GRAPHICS

ROUGH DRAFT

NOTES

OTHER

MY BLOG IDEAS
MIND MAP

MY BLOG IDEAS

SUBJECT

IDEAS

PICTURES/GRAPHICS

ROUGH DRAFT

NOTES

OTHER

MY BLOG IDEAS
MIND MAP

MY BLOG IDEAS

SUBJECT

_____
_____
_____
_____
_____
_____

IDEAS

_____
_____
_____
_____
_____
_____

PICTURES/GRAPHICS

ROUGH DRAFT

NOTES

_____
_____
_____
_____

OTHER

_____
_____
_____
_____

MY BLOG IDEAS
MIND MAP

MY BLOG IDEAS

SUBJECT

IDEAS

PICTURES/GRAPHICS

ROUGH DRAFT

NOTES

OTHER

www.ingramcontent.com/pod-product-compliance
Lightning Source LLC
Chambersburg PA
CBHW071300050326
40690CB00011B/2475